Ditching the Marshes

A History and Bibliography

Lyle Nyberg

Copyright © 2022 Lyle Nyberg. All rights reserved.

Published by Lyle Nyberg, www.lylenyberg.com
Scituate, Massachusetts, 2022

ISBN: 978-1-7354745-7-1 (paperback)
Also to be available in eBook/Kindle version

Cover: Aerial photo of Westport River salt marsh ditches behind Horseneck Beach, Westport, MA, by Gary M. Banks, William E. Richardson, 2022. Thanks to Susannah Green for assistance, and editors Janet Paraschos and Alix Stuart.

Ditching the Marshes

A History and Bibliography

Introduction

Here at Third Cliff in Scituate, Massachusetts, where the North River meets the Atlantic Ocean, we have salt marshes with ditches. And we have questions about those ditches. What are they? How old are they? What were they used for? Answers to these, and many more questions, are covered later in this document.

For now, the simple explanation is that ditches were dug in colonial times to mark property boundaries, and to drain and dry out marshes (at least at low tide) in order to facilitate farming of salt marsh hay. The hay was used to feed cattle and horses. Starting in the early 1900s, more ditches were dug to drain pools of standing water where mosquitoes bred. (Tides would still flow in and out of salt marsh ditches.) In addition, mosquitoes were attacked by applying oil to their breeding grounds, and applying chemicals like DDT. Ditching (meaning the act of digging ditches) expanded in the 1920s and 1930s in the war against mosquitoes by regional and state authorities, and federal agencies like the Civilian Conservation Corps and the Works Progress Administration. These ditches have been maintained up to present times.

From a modern perspective, ditches create more complex questions. For one, ditching seems to cause marshes to sink, leaving them more vulnerable to rising sea levels. To encourage growth of the marshes, some ditches are being remediated by filling them. Much work and research is percolating on these questions. So I naively wonder — to fill the ditches or not to fill — is that the question? Or is that putting it too simply?

I dug up lots of information about ditches. They are part of the local history of many places along North America's northeast shores, including Canada's Bay of Fundy; Hampton, NH; Ipswich, Newburyport, and Scituate in Massachusetts; and New Jersey. Some of that local information is listed below, but this just scratches the surface.

Second Cliff (top) and Peggotty Beach in 1968. Straight ditches (in sets of parallel lines) drain to tidal creeks in marsh. Source: Town of Scituate, Town Archives, CGL-134.

Information on ditches is scattered in time as well as in space. Early settlers brought with them the practices of their home countries. The English colonized Scituate, like many parts of New England. England has a long history of digging ditches, as noted below.

My research found so much information that it seemed best to share it in a kind of bibliography, as set out below. I try to let the sources speak for themselves. There is no particular order or style (although I tend to follow chronological order, and the Chicago Manual of Style for notes). I add many hotlinks, so an electronic version of this document may be most useful. The document is divided into sections:

For an overview of ditching, I highly recommend, in the third section, the OWMMS history and the NSRWA talk (Feb. 2022), both online. In the same section, the Sutter and Rudin materials are also excellent. But there is a wealth of further information below. Much of it comes from experts, which I am not. This is just intended as a guide, resource, and starting point.

Please let me know if this goes off-track, or if you have additional resources to suggest. Contact me at www.lylenyberg.com. My website has a Ditches tab where you can download a version of this bibliography. The website also has a downloadable version of my form on the North River Salt Marshes and Marsh Ditches submitted for the Massachusetts Cultural Resource Information System (MACRIS). It covers almost 2,000 acres of the Scituate portions of the marshes. As of this writing, it appears it will not be published in MACRIS because it goes beyond the purpose of the form.

I don't mind going "beyond," for several reasons. First, my research found no document like this, gathering sources on ditching. Well, except for that 1662 treatise on the ditches of England. Second, my town of Scituate (incorporated in 1636) has a significant history of salt marsh ditches. Deeds and documents referring to ditches are cited below. Third, the past ditching of salt marshes, whether to help foraging and farming of salt marsh hay, or to eliminate mosquitoes, may not be exactly what we want for the future. Included below are many studies of the ecology and value of marshes, and their threats. It is a puzzle in process, and I hope this helps provide some of the pieces to that puzzle.

Marshes are special places. I live near salt marshes. My early views fixed on the straight lines of the ditches. Perhaps we are naturally drawn to such incongruities in a landscape, things that stand out. Perhaps it is easier to contemplate the parts instead of the whole. But once in the marsh, it is difficult not to appreciate the whole scene — its colors, its openness (at least when adjacent to a river like the North River), and its exposure to the sky, unobstructed by trees or buildings.

Marshes are wide, horizontal landscapes with low horizons. They are like many of the Dutch paintings of the 1600s. They are another world. Yet they account for much of my town's remaining open space, and they are a significant buffer between the land and the rising seas. It is fair to say that my study of ditches has opened my eyes to the wonders of the marshes.

How did this happen? It's my incessant curiosity, and drive to understand the world around us. I wrote about my neighborhood in *On a Cliff*, and that partly explored the ditches around here. Then I wrote about our town's seacoast in *Seacoast Scituate By Air*, and got to see how our ditches and marshes look from above. It was only a natural step to go from up in the air to down in the ditch!

Lyle Nyberg
www.lylenyberg.com
Scituate, Massachusetts
August 21, 2022

1 Early History

Digging ditches goes far back in history. The English knew ditches. Some were dug for defensive purposes, as around a castle or outside the walls of London. Some were dug to delineate boundaries. Others were dug to improve marshy areas. The English brought their knowledge of ditches to the New World in settling New England. Many of those from Kent, England, settled Scituate, Massachusetts.

a. "In some world regions, societies undertook large-scale environmental transformations of watery landscapes in order to make them suitable for farming or herding. One prime example is that of the North Atlantic wetlands. Since Roman times, the marshlands of Europe had repeatedly been ditched and/or diked, to drain the wetlands and/or block the inflow of saltwater (Hatvany 2003). These tactics produced varied results. In the notorious marshlands of Kent, England, the diking of the marshlands in the sixteenth century may have inadvertently improved the habitat for *Anopheles atroparvus*, the most important malaria mosquito vector in the British Isles, and consequently increased malaria transmission. But by the second half of the nineteenth century, the incidence of malaria in these wetlands had dropped dramatically." James L.A. Webb, Jr., "The Long Arc of Mosquito Control," ch. 4 in Marcus Hall and Dan Tamïr, eds., *Mosquitopia: The Place of Pests in a Healthy World* (Milton Park, England: Routledge, 2021), 49-60, eBook at https://doi.org/10.4324/9781003056034-4. It's interesting that ditching encouraged mosquitoes, and was later used to eradicate them. It probably depended on the kinds of ditches or the various uses of the word; see the note below on terminology.

b. For a deep history of marshlands and ditching, by a historical geography professor, see Matthew George Hatvany, *Marshlands: Four Centuries of Environmental Change on the Shores of the St. Lawrence* (Sainte Foy, Canada: Les Presses de l'Université Laval, 2003); see https://books.google.com/books?id=-Hk9eDMLFFkC&source=gbs_navlinks_s, and https://archive.org/details/marshlandsfource0000hatv/page/n211/mode/2up. About as thoughtful a history as you can get. The book's lessons and insights apply to all North America, particularly the salt marshes of New England.

c. John Speed (1551–1629) was a famous English cartographer, chronologer, and historian, still notable today. His descriptions of Great Britain referred to ditches. One ditch marked the boundary between Suffolk and Kent. Others were believed to be the creation or home of the Devil. One "admirable Trench" ran for more than 100 miles, separating the Welsh from the English. John Speed, *England, Wales, Scotland and Ireland: Described and Abridged with Ye Historic Relation of Things Worthy Memory: from a Farr Larger Voulume* (London 1627, 1666, etc.), https://books.google.com/books?id=KeRBAQAAMAAJ&newbks=1&newbks_redir=0&source=gbs_navlinks_s. The "Farr Larger Voulume" is *The Theatre of The Empire of Great Britaine.*

d. The following description from 1662 tells the history of Moorfields, admittedly not a salt marsh, but a marshy area north of London (thought to have been created by the building of London's wall):

> This Fen, faith Stow [according to chronicler John Stow?], stretching from the wall of the City, … continued a waste, and an unprofitable ground a long time, …But in the year MCGCCxv [1315] Thomas Fawconer, Maior, caused the wall to be broken towards the Moor, and builded the Postern [secondary door], called Mooregatt, for ease of the Citizens … Moreover,

> he caused the Ditches of the City, … to be newly cast and clensed; by means wereof the said Fen or Moor, was greatly drained and dryed. [paragraph break] And in the year MDxij [1512], Roger Atchley, Maior, caused divers Dikes to be cast, and made, to drain the waters of the said Moore fields, with bridges arched over them; [and later works] caused divers Sluces to be made, to convey the said waters over the Town ditch, into the course of Walbrooke, and so into the Thames: and by these degrees was this Fen or Moore, at length made main and hard ground, which before, being overgrown with Flaggs, Sedges and Rushes, served to no use.

William Dugdale, *The History of Imbanking and Drayning of Divers Fenns and Marshes, Both in Forein Parts, and in This Kingdom; And of the Improvements thereby* (London, n.p., 1662), 73–74, https://books.google.com/books?id=bS6-fDCYwoYC&newbks=1&newbks_redir=0&source=gbs_navlinks_s. An amazing historical survey. Dugdale also wrote this, at page 184, evidently about events that occurred during the reign of Brithnod, Abbot of Ely, A.D. 970–981:

> [The Abbot of Thomey] calling together the Inhabitants on every part thereof, assigned them their several portions of land: and for a perpetual evidence of the possessions belonging to his Church, he caused that large and deep ditch to be cut through the main body of the Fenn, which was then called by the name of Abbot's deike (as I have observed) to the end it might remain as a boundary in that deep mud and water.

In addition, Dugdale's work shows that "ditch" and "dike" (based on the Old English word "dic") were sometimes used interchangeably. See note below on terminology. See also "Grim's Ditch," Wikipedia, https://en.wikipedia.org/wiki/Grim%27s_Ditch; and "New Ditch," Wikipedia, https://en.wikipedia.org/wiki/New_Ditch. Dugdale's work records jury decisions on boundary disputes relating to ditches.

e. Many English place names end in "ditch" — for example, *Shoreditch*, a district in London; *Houndsditch* surrounding London's wall (Chris Horne and Neil Adams, "City Ditch" in Janelle Jenstad, ed., *The Map of Early Modern London, Edition 7.0* (Victoria, CA: University of Victoria), https://mapoflondon.uvic.ca/DITC1.htm); and *Reddish* near Manchester (item 28 in Paul Anthony Jones, "From Ham to Sandwich: 40 Odd British Place Names (And What They Really Mean…)," Mental Floss, November 10, 2016, https://www.mentalfloss.com/article/88110/ham-sandwich-40-odd-british-place-name).

f. Turning to the New World, salt marshes in New England were highly valued by the early settlers because they provided salt marsh hay to feed their animals. The Boston Harbor islands were a source of hay, dating to 1634 or earlier. Nancy Seasholes, "Haying" Boston Harbor Islands National Recreation Area website, https://www.nps.gov/boha/learn/historyculture/haying.htm. "About this same time (1638) a plantation was formed at Winnicumet, which was called Hampton [NH]. The principal inducement to the making this settlement was the very extensive salt marsh, which was extremely valuable, as the uplands were not cultivated so as to produce a sufficiency of hay for the support of cattle." "[Hampton, NH's] Birthday," *Boston Globe*, August 15, 1888, 4.

g. "Scituate's great appeal to the early settlers was the vast acreage of marshland behind the cliffs which at that time extended much farther out into the sea. … Marsh hay assured the farmers of winter feed for their animals. The marshes were carefully conserved, and there were restrictions against allowing hogs on the marshes. … Cows were here at the time Scituate was settled, the first having arrived in Plymouth in 1624." Nathaniel Tilden, "Farming Scituate Marsh Lands," *Scituate Mirror*, May 8, 1975, 12, available on Scituate Town Library website. Good overview from a descendant of one of the original English settlers of Scituate.

h. Early Plymouth Colony records, up to about 1651, show that individuals were assigned parcels for mowing "hey" (hay). These were often described as "marsh," "mersh" (marsh), or "marsh meddow" (meadow). Some were salt marshes, some were fresh marshes. A word search for "marsh" yields many results. (The word "ditch" appears infrequently.) Nathaniel B. Shurtleff, ed., *Records of the Colony of New Plymouth in New England*: Printed by Order of the Legislature of the Commonwealth of Massachusetts, Vol. 1 & 2 (Boston, 1855), https://archive.org/details/recordsofcolonyo0102newp/page/n7/mode/2up; https://www.sec.state.ma.us/arc/arcdigitalrecords/pcolony.htm. For example:

   i. "Orders about mowing of Grasse for the p'nt Yeare, 1633" including "For Edw. Wynslow that against his own ground, & from the marsh over against Slowly House up the river." (14–15)

   ii. "At a Genall Meeting the xxth of March, 1636, according to the Order of the Court, these Hey Grounds were assigned to the Inhabitants of Plymouth, Eele River, & Ducksbury" including "To Thomas Cushman, the remainder of the marsh before the house he liveth in, (wch Mris Fuller doth not use,) and the little pcell at the wadeing place on thother side Joanes River." (55-57)

i. Salt meadows in the new world (1655): "There also are brooklands and fresh and salt meadows; some so extensive that the eye cannot oversee the same. Those are good for pasturage and hay, although the same are overflowed by the spring tides, particularly near the seaboard. These meadows resemble the low and out lands of the Netherlands. Most of them could be dyked and cultivated." Adrian Van der Donck, *Description of the New Netherlands* (1655, per Old South Leaflets No. 69, 1896), pamphlet, 13–14,

https://digitalcommons.cedarville.edu/cgi/viewcontent.cgi?article=1053&context=pamphlet_collection.

j. "There is towards the sea a large piece of low flat land which is overflowed at every tide, like the schorr with us, miry and muddy at the bottom, and which produces a species of hard salt grass or reed grass. Such a place they call *valey* and mow it for hay, which cattle would rather eat than fresh hay or grass. It is so hard that they cannot mow it with a common scythe, like ours, but must have the English scythe for the purpose." Jasper Danckaerts, *Journal of Jasper Danckaerts, 1679–1680* (New York: Charles Scribner's Sons, 1913), 60 (referring to Coney Island at southern end of Brooklyn, NY), https://archive.org/details/journalofjasper00danc/page/n11/mode/2up. In addition, at 97–98 (and see also 147): "The island [Matinnaconk, in the Delaware River, between Bordentown and Burlington, probably Burlington Island, NJ; see W. W. H. Davis, *The History of Bucks County, Pennsylvania, from the Discovery of the Delaware to the Present Time* (Doylestown, PA, 1876), 24, https://archive.org/details/historybuckscou00davigoog/page/n32/mode/2up?q=1657], formerly, belonged to the Dutch governor, who had made it a pleasure ground or garden, built good houses upon it, and sowed and planted it. He also dyked and cultivated a large piece of meadow or marsh, from which he gathered more grain than from any land which had been made from woodland into tillable land."

k. Draining wet grounds (1684): "But if this marsh and filthy ground do not lye so low as these low valleys, but rather against the tops of hills; you shall then, first open the heads of all the springs you can find, and by several drains and fluces, draw all the water into one drain, and so carry it away into some neighbouring ditch and valley; and these drains you shall make of a good depth, as at least two foot, or 2 foot and a half, or more, if need require, and then cross-wise every way overthwart the ground, you shall draw more shallow furrows, all which shall fall into the former deep drains, and so make the ground as constant, and firm as may be …." Gervase Markham, *Markham's Farewel to Husbandry* … or *A Way to get Wealth … The thirteenth time corrected, and augmented by the author, Volume 5* (London: Gervase Markham, 1684), book 2, 43, https://books.google.com/books?id=8idgsaA9T18C&newbks=1&newbks_redir=0&source=gbs_navlinks_s. Good advice, perhaps, for draining marshes.

l. Here we must pause for a note on terminology. Van der Donck (1655) and Danckaerts (1679) use the word "dyke" while Markham (1684) uses the word "ditch." The two words are connected, as Jeremy Bangs pointed out to me in a July 4, 2022, email from the Netherlands: "perhaps an ancient connection when Dutch and English were much closer to each other, ca. 700 years ago." Not only were the countries just across the channel from each other, but also the Pilgrims spent years in Leiden before traveling to Plymouth.

A "ditch" connotes a low channel, commonly used to drain water, as in Markham's example, but a channel can also receive and circulate water, as in an irrigation ditch. The vagueness in the word "ditch" allowed it to describe both a breeding ground for mosquitoes as well as a method for draining mosquito grounds, as noted above in the quote from Webb (2021).
As if to compound the confusion, a "dike" carried two meanings: "Dike, or Dyke…1. A channel to receive water. 2. A mound to keep water in its channel." J. and R. Fuller, *A General Dictionary of the English Language: Compiled with the Greatest Care from the Best Authors and Dictionaries Now Extant. By a Society of Gentlemen* (London: J. and R. Fuller, 1768), https://books.google.com/books?id=DQMaLg65mo0C&newbks=1&newbks_redir=0&source=gbs_navlinks_s. The first meaning seems equivalent to an irrigation ditch. But the second meaning is opposite the first, at least in height — a channel is low, a mound is high. Obviously, the material dug out of a ditch could be (and often was) banked aside of it, creating an embankment, wall, or dike. Today we are influenced by the 1865 American story of the Dutch boy who saves his country by putting his finger in a leaking dike (embankment). True, a ditch in a proper sense of the word could be used (with an embankment) to keep salt water out of upland farmed fields, or (less commonly) to keep fresh water in those fields. However, we use the term "ditch" in this document mostly to mean a channel to drain a marsh. By the way, a "ditch" is somewhat distinct from a "graff" or "guzzle" or "gutter." We have to be careful with words, particularly when covering various writers in separate centuries and different countries (England and the Netherlands).

m. William Cronon, *Changes in the Land: Indians, Colonists, and the Ecology of New England* (New York: Hill and Wang, 1983, first revised ed. 2003). A classic book. Discusses salt marshes and drainage, not ditches specifically.
    i. "As important as the shore itself were the salt marshes. Here the tides regularly flooded extensive inland areas with salt water, so that only two grasses — *Spartina patens* and *Spartina alterniflora* — were able to grow there. Because the grasses helped accumulate soil and so created a series of microenvironments from dry land to marsh to protected pools of water, they furnished a home for a wide variety of insects, fish, and birds." (Page 31.) The grasses were essential for the colonists since the new lands were heavily forested and were cleared to create farmland.
    ii. Colonists cleared trees, and, long term, this dried up the countryside and reduced water tables. 125–126.
    iii. "Toward the end of the eighteenth century, other colonial activities began to have significant effects on New England ecosystems. The draining of swamps and salt marshes became more frequent as greater amounts of capital were invested in agriculture." 155.

2 Early Records and Deeds

Early colonial records and deeds have many references to marshes. References to ditches do not show up until the latter part of the 1600s.

a. Jeremy Dupertuis Bangs, *The Seventeenth-Century Town Records of Scituate, Massachusetts*, three volumes (Boston: New England Historic Genealogical Society, 1997, 1999, 2001), https://catalog.hathitrust.org/Record/010029242. Astonishingly detailed and authoritative. Documents include digests of some land divisions and property deeds. They refer to marshes (many) and ditches (none in volume 3). Some are listed below, in rough chronological order, with citation to volume and page of Bangs' book.
   i. 1633 land grant with bounds including "the channel in the marsh to the west." This was probably a natural channel (stream) rather than a ditch that had been dug out. I:75, item 58.
   ii. 1636/7, February 7. Marshland grants. "On 7 February 1636/7, marshland on the east side of Kent Street was granted to Richard Foxwell, described as 'lying at the east end of this lot on the other side of the highway' (R. 102). Similarly, on the same date, marshland was granted to Walter Woodworth, 'lying on the east side of the highway against the end of his lot of upland' (R. 106). Also on that date a portion of marsh; and upland was granted to Samuel Hinckley 'adjoining together on the other side of the highway over against his house lot the breadth of it being answerable to the east end of his lot and so continued to the Third Cliff' (R. 112). … The foregoing descriptions show a continuation of five evidently traditional terms for roads both before and after the incorporation of Scituate as a town in October 1636. … Also not specific as to public or private ownership is the term 'causeway,' which was a way laid over a marsh. … Construction details of a way laid over a swamp are found in a contract of 11 December 1660, when John Palmer agreed 'to make A sufficient way through the swamp Below the ffoote way that now is: … the way to bee A sufficient Cart way in the place Aforesayd, to bee made with Loggs Eleaven foote Long for the Breadth of the way with A sufficient water Course ….'" III:17–18.
   iii. 1654, July 31. Deed by Gov. Bradford. Deed for "mersh and Swamp lying and being in Scittuate neare a brooke called Scittuate brooke [now Satuit Brook]; …" III:389, App. 5. The deed mentions "fresh" marshes, presumably distinct from salt marshes. As shown in the deed, by this time, the marshes had clearly been divided among multiple owners. See also II:54 ("[n.d.] Timothy Hatherly sold the thirtieth marsh lot").

Marshes were sometimes named for their owners. III:389, App. 5.; II:306 (1685, Pooles ditch).

iv. 1664. "Rotten Marsh Swamp" was a meadow in front of the house of the late Judge William Cushing (at Belle House Neck). I:87, item 153.

v. 1662 examples of boundaries of marshes along North River, including near "Beell House" [Belle House]. I:159, item 456.

vi. 1662 Resoulved White grant or deed to Isack Buck. Intricate "Marsh medow" boundaries involving North River tributaries. The deed mentions "straight lines" (ditches?), including at Bear Island [near Cushing property at Belle House Neck], "…being bounded on the North by one straight Line Crose the midle of the said Iland and medow from East to west and soe to the first hering Brook the northerly halfe being the Lands of Peter Collemore and one the south is bounded by another straight Line that Runneth from East to West along by the south side of the body of the said Iland as the stackes [hay stacks?] now stand till it com nere to the mouth of cartayne greate Creeke that Issueth forth into the said hering brooke by which Line it is devided from the medow lands that sum times was Mr. William Vassall and now in the posestion of John Cushen …." Also, a right of passage (including cattle) through former lands of Vassall, "Commonly used and to bee used unto the hey [hay, probably, see Bangs, III:22, but could be high] ways beyond the said Lands …." I:259–260, item 45.

vii. Sample of Index to Place Names; includes Gillson's Marsh, Great Marsh, Herring Brook Marsh, Herring Marsh, Little Fresh Marsh, Little Marsh, Long Marsh, New Found Marsh, New Harbor Marsh, New Marsh, North River Marsh. I:536–537.

viii. 1666, June 27, land transfer including "'all that meadow lying at the southeast End of the sd upland between it & the North River'; the other meadow parcel being two acres, bounded to the east to a ditch dividing it from the meadow of John Bryant, to the south to a ditch dividing it from the meadow of William Curtis, to the west to William Curtis's upland, and to the north to William Randall's upland." I:163. By this time, ditches were clearly markers for property boundaries, but they may have been dug to drain the marshes.

ix. 1681, July 26. William Curtis deed to Edward Wanton for land "bounded by the North River on the east, bounded to the south to Edward Wanton's meadow, to the west to William Randall's land, and to the north 'to a ditch that is dugge from ye River almost to the head that parts William Curtis & Edward Wanton'". I:184. The head would likely be the marsh's border with the upland.

x. 1692, June 1. Five acres of marsh meadow, "Little marsh" [evidently bounded to the north and east by Herring Brook], conveyed by John Williams to John Barker, and he conveyed a one-quarter interest to Thomas Pinson on November 19, 1696. I:433; see also I:257, 1666 & 1677.

xi. 1695, June 1. Two ditches used to describe land conveyed at Farm Neck. II:301.

xii. 1709, September 27. Town record of agreement among Samuel Barker, John Holbrook, and John Holbrook, Jr., that, "ye ditch adjoyneing to musquaschcut pond runing from Sd pond to Israel Cowings medow Should be ye devideing line betweene the Sd holbrook farme on which they dwell and the medow of ye Sd Samuel Barker. And at the Same time they made a devision of ye fence ye length of Sd ditch; which is about 58: Rods." II:526.

b. Examples of early deeds, some mentioning salt marsh and ditch:

i. Marshes mentioned but not ditches in book of Plymouth County recorded deeds to 1651. David Pulsifer, ed., *Records of the Colony of New Plymouth in New England*: Printed by Order of the Legislature of the Commonwealth of Massachusetts, Deeds, &c. Vol. 1 [12 in series] 1620–1651, (Boston, 1855), https://archives.lib.state.ma.us/handle/2452/802284.

ii. Timothy Hatherly to John Williams, Sr., and John Williams, Jr., deed, February 5, 1651, Plymouth Country Registry of Deeds, book 433, page 99, recorded June 27, 1877 (long delay!) (multiple parcels including marshes; includes Tilden Farm?), transcribed in Jeremy Bangs, *Seventeenth Century Records*, III:552–553.

iii. Mathias Briggs to John Cushing, deed, October 11, 1670 (October 16, 1688?), recorded January 25, 1697, Plymouth County Registry of Deeds (online), book 3, page 8 (see transcribed image at PCRD website). Conveys interest in "houses lands salt marshes and meadows which were formerly the said Mr. William Vassalls" in Scituate, north of Belle House (Belle House Neck) along North River.

iv. Benjamin Woodworth to John Cushing, deed, June 23, 1691, recorded January 25, 1697, PCRD, book 3, page 11. Conveys title in house "and twenty two Acres of upland more or less lying and being in ye Town of Situate aforesaid which I now Dwell in and upon And a right of commons or undivided lands of said Town And also meadow marsh land and swamp land and adjoyning to said upland." Bounds of seven-acre marsh included "on Northeast and southeast with the marsh land of said Otis on the ?? and Ditch now standeth And runeth Down to said Herring Brooke."

c. In 1876, Scituate conveyed by deed certain salt marshes to South Scituate (now the town of Norwell). As described in WPA index cards at Town Archives (subject "Salt Meadow or North River Flats"), they were "The Gulph Island, The Middle Green Island & Sunken Flat, The Jacob Flat, The Northey Flat, The Great Green Island, The Great Flat, and The Hummock Flats — definitely bounded."

3 History/General

This section covers descriptions of salt marsh haying, then turns to mosquito ditching.

a. Henry Follensbee Long, "The Salt Marshes of the Massachusetts Coast," Essex Institute Historical Collections, Vol. XLVII, 1911, 5–19, https://archive.org/details/essexinstitutehi47esseuoft/page/n9/mode/2up: "And is there in Massachusetts a landscape or a touch of Nature's hand that carries the mind so far toward the creative period of the earth as to stand in the midst of the salt marsh when the flush of twilight steals over its straw-colored desolation. It is as if our eyes opened for the first time upon the struggle of darkness with light; as if chaos ruled again and the drama of evolution had to be created anew, for the salt meadows, with their lagoons and pools of standing water, are the last remnants of the sea which once covered them entirely." In the 1660s, marshland in Newbury, Massachusetts, was divided and laid out for the use (if not ownership) of various persons. This was typical in Scituate, as well. Long's discussion may be the definitive account of salt marsh haying. Thanks to Historic Ipswich for finding this.

b. Long, "Salt Marshes," page 8: "A marsh to produce good hay and to be in good condition to cut must be well drained, and these drainage ditches, cut with turf spades, are a source of constant danger to people unfamiliar with the marshes, and many a gunner or green marsh hand can relate a personal experience of having slipped into one of these narrow ditches three or four feet deep."

c. Scott W. Nixon, *The Ecology of New England High Salt Marshes: A Community Profile* (Washington, DC: Fish and Wildlife Service, U.S. Dept. of the Interior, March 1982), https://catalog.hathitrust.org/Record/012213000. Incredibly important and revealing study of salt marshes, including their history in New England. Chapter 5 "The Human Impact on the High Marsh" is particularly useful. Thanks to Jim Glinski for finding this.

d. James H. Allyn, Major John Mason's GREAT ISLAND (Mystic, CT: Roy N. Bohlander, 1976, second printing 1988), available from Mystic River Historical Society website, http://www.mystichistory.org/digital_publications/Major_John_Masons_GR

EAT_ISLAND.pdf. Describes marsh ditches dug by early English settlers, evidently to facilitate salt marsh haying, in an area near Stonington, CT. See pages 23, 26, 47 (late 1700s), 58, 83 (mosquito control), and maps 3 and 4 at beginning and/or facing pages 20 and 26.

e. Peter Vandermark, "Marsh Elder," *Boston Globe*, October 3, 1993, 302. Wonderful, detailed story of a modern-day salt marsh hay farmer, Dan McHugh, in Rowley, MA. The operation he founded at Great Meadow Farm says "Our wetland construction and restoration services are supported by the only salt-water nursery in New England. Great Meadow Farm's on-site wetlands nursery has a capacity of over 1.5 million plants." http://www.marshmadness.org/GreatMeadowFarm.html.

f. Heather Alterisio, "Newbury Conservation Commission moves forward with Green Meadow lease agreement," *The Daily News* [Newburyport, MA], February 17, 2021, updated March 2, 2021, https://www.newburyportnews.com/news/local_news/newbury-conservation-commission-moves-forward-with-green-meadow-lease-agreement/article_b2377dc0-1ffa-572a-8e56-59b63fc07edf.html. Salt haying continues at Great Meadow Farm on land owned by the town of Newbury next to the Parker River.

g. Paul S. **Sutter**, ch. 11 "The Environment," in Stephen J. Whitfield, ed., *A Companion to 20th-Century America* (Malden, MA: Blackwell Publishing Ltd, 2004), 181–182, https://ur.booksc.me/book/78401565/21ad42, DOI 10.1002/9780470998533.ch11 — "In 1780, what became the United States had more than 220 million acres of wetlands. Today there are only about 100 million acres left, and about half of that loss occurred during the past century (Vileisis 1997). Farmers traditionally have been the parties responsible for wetlands drainage; and they continued as major players in the twentieth century, though with new tools and support. Trenching machines made swift work of ditching, and government programs — drainage districts in particular — provided crucial assistance. Farmers were joined by federal agencies such as the Army Corps of Engineers, and by timber companies and real estate developers. Among the hardest hit areas were the prairie wetlands of the Midwest, southern bottomland forests, coastal marshes, California's Central Valley, and the Everglades. Public health concerns motivated some of this drainage early in the century, when malaria remained a problem in California and the South, but most drainage aimed to make what were considered wastelands useful agriculturally. Wetlands decline has meant the loss of rich wildlife habitat and the important ecological services that wetlands provide: from filtering runoff to buffering against flooding and coastal erosion (Prince 1997; Vileisis 1997; McNeill 2000; Rome 2001). Wetlands drainage deserves to stand with dam building and irrigation development in the history of large-scale water manipulation in the past century."

h. Massachusetts Open Marsh Water Management Workgroup, "Mosquito Control Open Water Marsh Management Standards" (2010) ("**OWMMS**"), at website of State Reclamation and Mosquito Control Board (SRMCB), https://www.mass.gov/state-reclamation-and-mosquito-control-board-srmcb. OWMMS Appendix A: "History of Salt Marsh Management for Mosquito Control in Coastal Massachusetts" is an excellent history of ditching.
i. Mosquito control summary, including history of MA State Reclamation and Mosquito Control Board (SRMCB), in part II. Introduction, https://www.mass.gov/doc/introduction-history-current-organization-and-practice-of-mosquito-control-in-massachusetts/download.
j. Plymouth County [MA] Mosquito Control Project (founded 1957), https://www.plymouthmosquito.org/. Habitat Reduction: "1. Maintenance of existing drainage. … 2. Open Marsh Water Management is the selective creation of pond and ditch systems on salt marsh. New work in this program is tightly regulated by the Army Corps of Engineers and MA Department of Environmental Protection. The goal of OMWM is to reduce mosquito production by increasing fish access to places that produce mosquitoes." The project has compact excavators modified to operate on marshes.
k. Mass Audubon and North and South Rivers Watershed Association (**NSRWA**), "Salt Marshes of the South Shore," video talk, February 2022, https://www.youtube.com/watch?v=9tgBCKSM33s. Susan C. Adamowicz finds subtle clues in the landscape that show how people shaped the salt marshes, and she discusses ditch remediation to keep marshes from sinking and to protect them against sea level rise. Shows printed advice from the 1800s for salt marsh changes to improve farming.
l. North River Commission (established 1978), http://www.northrivercommission.net/. Protects corridor of North River, the state's only Scenic Protected River. Has 1978 maps.
m. North and South Rivers Watershed Association (founded 1970), https://www.nsrwa.org//. NSRWA website has many useful materials.
n. US Fish & Wildlife Service, Rachel Carson National Wildlife Refuge, https://www.fws.gov/refuge/rachel-carson, "Rachel Carson National Wildlife Refuge was established in 1966 in cooperation with the State of Maine to protect valuable salt marshes and estuaries for migratory birds." Consider the salt marsh sparrow.
o. Wendall Waters, "Coastal ecologist says handling of piping plovers is a major Massachusetts success story," WickedLocal.com, June 15, 2022, https://www.wickedlocal.com/story/regional/2022/06/15/piping-plover-nesting-season-massachusetts-beaches-balance-conservation-recreation-

ecologist/7457507001/. Piping plovers nest in beaches near salt marshes, including at The Spit in Scituate.

p. Lyle Nyberg, *On a Cliff: A History of Third Cliff in Scituate, Massachusetts* (Scituate: by author, 2021), pages 83–88 (brief discussion of ditches, with photos). Associated website, www.lylenyberg.com, with versions of this bibliography.

q. John R. Stilgoe, *Shallow-Water Dictionary: A Grounding in Estuary English* (Princeton, NJ: Princeton Architectural Press, 2004, and earlier editions).

r. Alyson L. Eberhardt and David M. Burdick, "Hampton-Seabrook Estuary Restoration Compendium" (Durham, NH: PREP Reports & Publications, or National Oceanic and Atmospheric Administration Restoration Center, 2009), https://scholars.unh.edu/prep/102. Excellent work.

    i. Page 5: "Although the European settlers highly valued the salt marsh, their use of it was not without impacts. Heavy pasture use resulted in the high density of ditching that is still seen in the marsh today. By cutting ditches to drain the marsh, it was thought to sustain the abundance and improve the quality of the salt hay as well as increase the abundance of black grass (*Juncus gerardii*), another valuable marsh grass used as feed. Salt pannes were reduced in these systems not only by cutting ditches to connect them to the drainage network, but also because the dredge spoil from ditching was deposited in the salt pannes to reclaim them as high marsh."

    ii. Page 7: "In the late 1930s additional ditches were cut into the marsh to drain it of potential mosquito breeding habitat, further altering the marsh drainage patterns, vegetation and density of pannes. Unfortunately, the small fish that preyed on the mosquitoes lost their habitat as well."

s. Joseph Dow, *History of the Town of Hampton, New Hampshire: From Its Settlement in 1638, to the Autumn of 1892*, vol. 1 of 2 volumes (Salem, MA: Lucy E. Dow, 1898), 77–79, https://books.google.com/books?id=cYRDAQAAMAAJ&source=gbs_navlinks_s. In 1671, the town voted to penalize persons whose cattle (including horses) roamed and foraged the meadows and saltmarshes (commonly in winter), including haystacks standing there.

t. Karen Raynes, "Salt marsh haying was once a community event at Hampton Beach," Seacoastonline, September 22, 2020, https://www.seacoastonline.com/story/news/2020/09/22/salt-marsh-haying-was-once-community-event-at-hampton-beach/42679111/. Has postcard image of haystacks on staddles (pilings) to raise hay above high tide. Good in-depth history.

u. Botan Anderson, "The Great Marsh Haystacks of New England," One Scythe Revolution website, posted August 5, 2013, https://onescytherevolution.com/blog/the-salt-marsh-haystacks-of-new-england. Great description of salt marsh haying, with images, including Martin Johnson Heade, "Haystacks on the Newburyport Marshes."

v. Historic Ipswich, "Gathering salt marsh hay," February 12, 2021, https://historicipswich.org/2021/02/12/salt-marsh-hay/. An exceptionally detailed description with photos of salt marsh haying (no mention of ditches). Wonderful bibliography.

w. Warren [RI] Land Conservation Land Trust, "Salt Marsh Haying on the Palmer River," https://warrenlct.org/salt-marsh-haying-on-the-palmer-river/. Has illustration of salt marsh hay being loaded from Wallace P. Stanley, *Our Week Afloat* (1888).

x. Chris Haight, "A Graduate Student in the Marsh" website, March 2, 2013, https://saltmarshblog.wordpress.com/2013/03/02/salt-marsh-haying/. History, and this: "Repeated clearing keeps Spartina alterniflora from encroaching on Spartina patens communities."

y. Elizabeth Holden, et al, "Traditional Uses of the Great Marsh: A Review of Lesser-Known Resources in Massachusetts' Great Salt Marsh" (Medford, MA: Tufts University, Spring 2013), https://static1.squarespace.com/static/52769555e4b0c10d4aa24eb5/t/52f07422e4b080c60b22db8f/1391490081999/GM_Final+Report.pdf. Comprehensive overview, discussion of salt marsh haying with its ditching, and impressive list of references. Great resource.

z. Shaun Roache, US Fish & Wildlife Service, "Salt Marsh Haying," talk on September 30, 2020, https://lymeline.com/2020/09/old-lyme-historical-soc-duck-river-garden-club-present-zoom-program-on-salt-marsh-haying-tonight/.

aa. Lauren Healey, "Marsh Restoration as Climate Change Adaptation," July 10, 2020, https://www.gulfofmaineinstitute.org/single-post/2020/07/10/Marsh-Restoration-as-Climate-Change-Adaptation. Brief history of mosquito ditching.

bb. US Fish & Wildlife Service, notice of Ecological Restoration Project, May 10, 2021, https://eeaonline.eea.state.ma.us/EEA/emepa/mepadocs/2021/051021em/pn/Eco%20Rest%20NEWBURY.pdf. "The purpose of the project is to restore approximately 109 acres of salt marsh which have been heavily altered by historic salt marsh haying (ditching and berms) and Open Marsh Water Management (ditching and ditch plugs). These legacy infrastructures are now causing increased and prolonged inundation of the marsh surface, conversion of high marsh to low marsh, and vegetation dieback."

cc. Joseph Foster Merritt, *Old Time Anecdotes of the North River and the South Shore* (Rockland, MA: Rockland Standard Publishing Company, 1928), particularly "Gundalow Days on North River" chapter all about salt marsh haying, 21–27, https://hdl.handle.net/2027/wu.89077200764?urlappend=%3Bseq=90%3Bownerid=13510798901256997-94.

dd. Joe Chetwynd, The Humble North River Gundalow," North River Packet (Norwell Historical Society, December 2019), 1, https://img1.wsimg.com/blobby/go/3564ec8d-06a8-4768-8bd3-7a892ae6e046/downloads/December%202019%20newsletter.pdf?ver=1626632547747. Includes photo of gundalow carrying salt marsh hay on the North Shore.

ee. John M. Holman, "The Salt Marsh and Salt Haying," Scythe Supply website, article published in 2006, https://scythesupply.com/the-salt-marsh-and-salt-haying.html. Includes great old photo of farmers building a haystack.

ff. Kezia Bacon, "Salt Haying on the North and South Rivers," NSRWA website, February 2017, posted April 22, 2008, https://www.nsrwa.org/salt-haying-on-the-north-and-south-rivers/. Nice review. I question the statement, "Todays ditches, created largely by WPA workers in the 1930's …" because it seemed like state and local agencies carried out much of the work.

gg. John Teal, Mildred Teal, *Life and Death of the Salt Marsh* (Boston: Little, Brown & Company, 1969, and later editions). This is what it would look like if a marsh could write its own biography. Very readable and comprehensive in scope. Covers nearly all activities affecting the marsh, including mosquito eradication and marsh ditching. Suggests that early ditches were dug to make navigation among the curving creeks faster and easier (page 30), and later ditches to drain and dry the marshes for haying (page 42). Examines digging and filling of Boston's marshes starting in 1641 (page 241). Addresses threats to marshes in chapter 16 (page 257), not emphasizing risk of sea level rise; otherwise, raises concerns that are still current years after this book was published.

hh. William Gould Vinal [also known as "Cap'n Bill"], *Salt Haying in the North River Valley (1648–1898)* (Cohasset, MA: by author, 1953), Scituate Town Library and Scituate Historical Society.

   i. Page 6 summarizes some deeds from 1788 on, first mentioning a ditch in a 1796 deed, conveying land "with the privilege of a way to cart the hay of the growth of said salt meadow over the land of Colonel John Cushing …"
   ii. Discussion of salt haying based on the account books of two Revolutionary War soldiers. Page 14.
   iii. "Salt haying took place after the garden crops were well underway and the English hay was on the mow. Winter and spring work consisted of carting salt hay to tide over the diminishing supply until pasture time. The one spring day in 1794 that William Barrell worked on the flats might have been 'ditching' as that was the only off-season work that was needed. There is no record as to when ditches were first made." 16.
   iv. Page 20 references Dr. Henry F. Howe, *Salt Rivers of the Massachusetts Shore* (1951), on file in Scituate Town Library. Page 102 of Howe's book

says, "Soon the thriving town [Scituate] began to spread inland along the tidal marshes of the North River. … By 1640 all the shore from Plymouth to Cohasset was filled with little farm communities. They all adjoined the salt marsh areas, where bedding and hay for their increasing herds of cattle were available."

v. Around page 22 — "excursion to a salt meadow in the gay nineties was a major event." Went down Neal Gate Street (at Belle House Neck) to the marsh.

vi. "Bunching, poling, tedding, raking scatters, pitching, stomping the load, and warily plodding home are obsolete performances. This vocabulary has stolen away like Arabs in the night. The words haymaker, pitching and heckling, have taken on new meanings." 24.

vii. Page 25 discusses ditches, based on an interview with an 89-year-old resident (so, born about 1860) and memories of his grandfather. "The line ditches marked boundaries. These were necessary to prevent the indiscriminate cutting of hay. The side ditches came in at right angles to help drainage. The meadow lands were marked off into rectangles because this was an old English custom."

viii. Page 26: "staddles" to stack hay appeared in Barnstable and Medford, not Scituate.

ii. In 1918, Scituate voted to contribute funds to "better the draining conditions on the marshes at North Scituate Beach." WPA index cards at Town Archives (subject "Marshes, Drainage of").

jj. Elizabeth Aykroyd and Betty Moore, *Hampton and Hampton Beach*, Postcard History Series (Charleston, SC: Arcadia Publishing, 2005) (not Grace C. Lyons title of 2016), 12–13, https://books.google.com/books?id=T20BzTSUFicC&printsec=frontcover#v=onepage&q&f=false. Good photos and brief history of haystacks in marshes of Hampton, NH.

kk. Daniel Wolff & Dorothy Peteet, "Why A Marsh," *Places Journal*, May 2022. https://placesjournal.org/article/the-deep-history-and-uncertain-future-of-a-marsh-on-the-hudson/. Outstanding.

ll. Ronald **Rudin**, *Against the Tides: Reshaping Landscape and Community in Canada's Maritime Marshlands* (Vancouver, BC: UBC Press, August 2022). "*Against the Tides* is the never-before-told story of the Maritime Marshland Rehabilitation Administration, a federal agency created in 1948 to reshape the landscape in the Bay of Fundy region." https://www.ubcpress.ca/media/UBCP_F22_CDN_forWeb.pdf; preview at https://www.amazon.com/Against-Tides-Reshaping-Landscape-Marshlands-ebook/dp/B09LRN19KY/. Ditching practices in England carried over into Canada's Bay of Fundy region in 1760s, with marshes used mainly to produce hay (pages 40–42; see also Foreword). The book is accompanied by a film, *Unnatural Landscapes*, http://www.unnaturallandscapes.ca/.

mm. Kimberly R. Sebold, "From Marsh to Farm: The Landscape Transformation of Coastal New Jersey" (Washington, DC: National Park Service, 1992), https://irma.nps.gov/DataStore/DownloadFile/484754, and https://archive.org/details/frommarshtofarml00sebo. Incredibly detailed and scholarly study of salt marsh farming and ditching practices. Well-illustrated. Describes and shows special mud-boots (meadow boots), like snowshoes, for horses to work on the meadows; also barge-like hay scows (gundalows) for loading hay; also dikes and embankments.

nn. Kimberly R. Sebold, "Low Green Prairies of the Sea: Economic Usage and Cultural Construction of the Gulf of Maine Salt Marshes," PhD diss. (University of Maine, 1998), available from author, kimberly.sebold@maine.edu.

4 Paintings

Marshes were a favorite subject for painters in the late 1800s and early 1900s.

a. Kimberly R. Sebold, "'Amid the Great Sea Meadows': Re-Constructing the Salt-Marsh Landscape through Art and Literature," *Maine History*, vol. 40, no. 1, article 4 (March 1, 2001), https://digitalcommons.library.umaine.edu/cgi/viewcontent.cgi?article=1212&context=mainehistoryjournal. Nice description and images of salt marshes and salt marsh haying, no discussion of ditches.

b. Fitz Henry Lane (1804-1865) painted nautical subjects, and occasional marsh scenes (without ditches), such as "Annisquam Marshes near Gloucester, Massachusetts" (1848), and "Pretty Marsh, Mt. Desert Island" (c.1850). Copies of many of his works are collected here: http://fitzhenrylaneonline.org/index.php. Many of his works are held by the Cape Ann Museum, https://www.capeannmuseum.org/.

c. Martin Johnson Heade (1819–1904), known for his salt marsh landscapes, seascapes, etc., including "Newburyport Meadows" (c.1876–1881) (salt hay stacks on staddles), Metropolitan Museum of Art, https://www.metmuseum.org/art/collection/search/11053; and "Sunset on the Marshes," (1887), "Massachusetts Modernist and Marsh-Themed Works Form 2 New Exhibitions at Cape Ann Museum," February 20, 2022, https://www.artfixdaily.com/news_feed/2022/02/20/5027-massachusetts-modernist-and-marsh-themed-works-form-2-new-exhibit; and Joey Ciaramitaro, "Window on the Marsh Celebrates Paintings, Photographs Capturing Great Marsh's Distinct Beauty," Good Morning Gloucester, posted March 16, 2022, https://goodmorninggloucester.com/2022/03/16/window-on-the-marsh-celebrates-paintings-photographs-capturing-great-marshs-distinct-beauty/

d. Frank Thurlo (1828–1913). See Virginia Bohlin, "Art, nostalgia reign in Newburyport show," *Boston Globe*, October 9, 1983, 91.
e. Arthur Wesley Dow (1857–1922), including "Ipswich Marsh 1900," "Flood Tide in the Ipswich Marshes," and "Haystack in the Marsh," at "Arthur Wesley Dow," February 10, 2022, Historic Ipswich website, https://historicipswich.org/2022/02/10/arthur-wesley-dow/; and Arthur Wesley Dow, "The Derelict, or The Lost Boat," woodcut (haystacks in the marshes), 1916, Metropolitan Museum of Art, New York City, Britannica, https://www.britannica.com/biography/Arthur-Wesley-Dow.
f. Thomas Buford Meteyard (1865–1928) — salt marshes in Scituate, late 1800s.
    i. See Mark Murray Fine Paintings, https://www.markmurray.com/thomas-buford-meteyard-paintings-for-sale.
    ii. "Winter, Scituate, Massachusetts, U. S. A.", 1890s, shows haystacks, perhaps at North River estuary with Colman Hills in background, and possibly the James House (now Maritime & Irish Mossing Museum) at right, Pallant House Gallery, https://artuk.org/discover/artworks/winter-scituate-massachusetts-u-s-a-70617/search/keyword:meteyard--referrer:global-search.
    iii. "The Beach, Scituate," showing beach connecting Third Cliff and Fourth Cliff before 1898 storm destroyed it, with North River marshes to the right (no ditches visible), Indianapolis Museum of Art, http://collection.imamuseum.org/artwork/55877/.
    iv. "North River, Moonlight, Scituate, 1893," with ditches, http://www.artnet.com/artists/thomas-buford-meteyard/north-river-moonlight-scituate-lrD562cUNQ_cDV1Z7tjLuA2l, originally exhibited at Berry-Hill galleries, and — with "Scituate, North River" (c.1894) — copied in *Scenes from Vagabondia: Thomas Buford Meteyard & Dawson Dawson-Watson: From Giverny to Scituate, 1890–1910*, exhibition catalog (Cambridge, MA: Pierre Menard Gallery, 2009).
g. The Rockport Art Association & Museum had an extensive exhibit in 2021, "Artists for the Great Marsh," with art by contemporary and past artists. https://www.rockportartassn.org/artists-for-the-great-marsh.

5 Photos and Maps

Early maps (1600s through 1800s) of the Massachusetts coast do not show ditches in marshes, as far as I could tell. Some show marshes and marsh creeks. Mapmakers would have avoided this level of detail and emphasized major natural and manmade features. Any early ditches might have been obliterated by shoreline development of marshes. The Boston shoreline, for example, was developed quite early (from 1641) for commerce and navigation, with piers and jetties that covered former marshland. See Nancy S. Seasholes, ed., *The Atlas of Boston History* (Chicago: The University of Chicago Press, 2019); Joseph G. Garver, *Surveying the Shore: Historic Maps of Coastal Massachusetts, 1600–1930* (Beverly, MA: Commonwealth Editions, 2006), esp. plates 12 and 14 (pages 26 and 30).

a. Early maps of Scituate, including from 1776, 1795 (two maps), and 1870 (one map), are cited and shown in Lyle Nyberg, Gary Banks, Bill Richardson, *Seacoast Scituate By Air* (Scituate: Lyle Nyberg, 2022). The book also has recent aerial photos of North River marshes and ditches.
b. O. H. Tittmann, H. L. Whiting, et al, US Coast Survey, *Sketch of North River, Mass.*, [map], scale 1:40,000 (n.p., 1870), Digital Commons at Salem State University, http://digitalcommons.salemstate.edu/maps_massachusetts/4/.
c. Ditches are not shown on most maps of the Scituate area. They are not on the USGS maps (1888, 1893, 1920, 1935, 1940), or US Coast Survey maps that I consulted, or the 1915 map of the North River at the North River Commission. They are not on the excellent 1903 map of Scituate. *Town of Scituate* [map], scale 1 inch = 1700 feet, in J. E. Judson, *Topographical Atlas of Surveys: Plymouth County together with the town of Cohasset, Norfolk County, Massachusetts* (Springfield, MA: L. J. Richards & Co., 1903), plate 31, including insets for Village of Scituate, North Scituate Beach, etc. ("1903 map"), State Library of Massachusetts, Massachusetts Real Estate Atlas Digitization Project, URI http://hdl.handle.net/2452/206055, http://www.mass.gov/anf/research-and-tech/oversight-agencies/lib/massachusetts-real-estate-atlases.html, and https://www.flickr.com/photos/mastatelibrary/9466953246/in/album-72157634981171273/. They are not on a 1906 plan of Rivermoor, Third Cliff.
d. Late 1800s and early 1900s photos and postcards show marshes, not ditches
    i. Late 1800s, "Haystacks on Hampton Falls and Seabrook Marsh, late 1800s," New Hampshire Historical Society, in Jerald E. Brown, *The Years of the Life of Samuel Lane, 1718–1806: A New Hampshire Man and His World* (Hanover, NH: University Press of New England, 2000), xix, https://books.google.com/books?id=a4bKRdX_6cEC&source=gbs_navlinks_s.
    ii. c.1905, Hampton Beach (639 x 327 pixels), https://en.wikipedia.org/wiki/File:The_Marshes,_Hampton_Beach,_NH.jpg
    iii. Samuel Peter Rolt Triscott, "Haystacks On The Salt Marshes At Hampton, New Hampshire," date unknown, hand painted oil reproduction, https://www.niceartgallery.com/Samuel-Peter-Rolt-

Triscott/Haystacks-On-The-Salt-Marshes-At-Hampton%2C-New-Hampshire-oil-painting.html.

iv. "Bird Conservation in the Hampton-Seabrook Estuary," NH Audubon brochure, c.2008, on Hampton marshes, https://www.nhaudubon.org/wp-content/uploads/Hampton-Brochure-final.pdf/

v. Botan Anderson, "The Great Marsh Haystacks of New England," One Scythe Revolution website, posted August 5, 2013, https://onescytherevolution.com/blog/the-salt-marsh-haystacks-of-new-england.

vi. "The Marshes, Plum Island, Newburyport, Mass." Postcard, Historic New England, https://www.historicnewengland.org/explore/collections-access/capobject/?refd=PC001.03.01.TMP.037.

vii. "Salt haying, Polpis, Nantucket," postcard, date unknown, a farmer collects cut hay in an ox cart in a salt marsh, Historic New England, https://www.historicnewengland.org/explore/collections-access/capobject/?refd=PC001.03.01.TMP.043.

e. Aerial photos are a good way to show ditches. For 1950 and later, the MacConnell collection at UMass Amherst has aerial photos of Massachusetts, online, https://credo.library.umass.edu/view/collection/mufs190. The collection is a jumble, but there is a map that lets you home in on an area, at least for those from 1950 to 1952. See more guidance at the bottom of this page on my website — https://www.lylenyberg.com/copy-of-3d-cliff.

f. Bound set of 30–40 large printed photos in black and white, covering the whole town, taken about 1954–1961, Town of Scituate, Town Archives. Some cover parts of North and South Rivers.

g. 1968 aerial photos, large black and white interpositives, covering the whole town, Town of Scituate, Town Archives. A few are copied in Nyberg, *On a Cliff.*

h. 1978 North River Commission ortho (aerial) maps, downloadable, http://www.northrivercommission.net/Maps.php.

i. Various shoreline studies for towns include aerial photos.

j. www.HistoricAerials.com

k. Salt marsh ditches of the Northeast are mapped here: Northeast Conservation Planning Atlas, "Salt Marsh Ditches, Version 3.0, North Atlantic U.S. Coast," Data Basin website, posted March 12, 2019, https://databasin.org/datasets/8b04346487a8457d98feea228d9b0275/.

l. 1922. Ditches are shown on the original Town of Scituate assessor maps of July 17, 1922. There are not many, but some appear at the southern end of Third Cliff. (Map 64, covering a small part of the North River marshes.) They appear

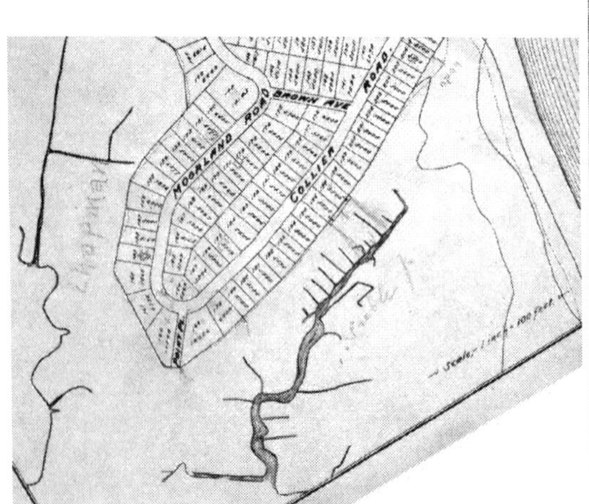

1922 Assessor's Map 64 (rotated)

1952 MacConnell aerial photo, DPT-5K-107

1968 Town of Scituate aerial photo, CGL-136[?] (detail, edited)

2020, Oct. 18, Google Earth

Left: Index map (detail) for 1968 Scituate aerial photos. Those showing North River marshes would include CGL-12, -14, -175, -177, -136, -137, -127, and -139.

as parallel lines in the marsh, echoing street lines in the Rivermoor development there. The map might not show all ditches there at the time. A 2016 edition of the map (in the lobby of the Town Hall) was still dated 1922 and nothing had changed, except that on the land, house plans and certain other informational items have been added to the house lots. (Compare portions of the 1922 plan to the 1968 aerial photos, which were taken for the town and are available in the Town Archives.) The nine parallel ditches on the right (east) side of Third Cliff are/were about 90 feet apart, a distance consistent with later ditching methods (WPA ditches were 115 to 230 feet apart, per Nixon 1982). They appear in the 1968 aerial photo, but there are additional ditches. One is a long, straight transverse ditch that does not seem to appear on the 1952 MacConnell aerial photo. The transverse ditch connects the nine ditches and runs parallel to the shore and close to the shore. A sand bar partly covers the nine parallel ditches. Today, the sand bar has shifted further west, covering most of the nine ditches, but the transverse ditch remains, as can be seen in a 2020 Google Earth photo.

m. The USDA has historical aerial photos, back to 1965 or earlier. https://www.fpacbc.usda.gov/geo/index.html. Plymouth County, MA, is designated 25023 (DPT). The photos are from the National Agriculture Imagery Program (NAIP), https://naip-usdaonline.hub.arcgis.com/, and predecessor programs like NAPP, NHAP, FSA, and NRCS.

6 Early 1900s Ditching

We think of mosquito ditching as a practice of the 1920s and 1930s. Surprisingly, however, it really became popular by the early 1900s. Shore towns did not want mosquitoes to discourage people from visiting or vacationing at their beaches. The year 1900 marks a change in the focus of ditching marshes, from harvesting salt marsh hay to eliminating mosquitoes. Following is a chronological selection of articles from the *Boston Globe* in the early 1900s. The first article involves land development in Indiana. The rest deal with mosquito ditching.

a. "Large Indiana Farms; One of 32,000 Acres Was Mostly Marsh Land a Few Years Ago — Its Ditches Are as Big as Small Rivers," *Boston Globe*, February 11, 1900, 29. "Instead of fences Mr Gifford has waterways between pastures. He has 75 miles of large ditches through the farm and has thousands of miles of smaller ditches."
b. "Found Dead in his Room; Mark B. True, Newburyport Inventor, Passes Away at Hotel," *Boston Globe*, November 10, 1902, 14. "At the time of his death he was experimenting with an improved marsh land ditching machine and a trolley device …"

Illustration for "Reclaiming West Island From The Mosquito," *Boston Globe*, May 30, 1909, 40 (West Island at Fairhaven, MA)

c. "Mosquito Fight. Board of Health Won at Lawrence. Dug Ditches and Opened Breeding Pools to the Tides. Then Residents Could Sit Out On Lawn." *Boston Globe*, April 12, 1903, 36. The Lawrence, MA, board of health reported, "For the disinfection of the salt marshes the petroleum method must be condemned as an expensive makeshift … It is too costly to compete with ditching and filling, which permanently dispose of the breeding places. The aim of the ditching process is to open every pool and pocket along the edge of the uplands to the free influx of every tide. [Cites successes at Isle of Wight.]"

d. "Reclaiming West Island From the Mosquito; Ditching and Draining the Marshes Destroys the Breeding Places of a Pest Which Has Always Prevented Man from Enjoying the Sea Breezes of Sconticut Neck — Draining is Made Possible By the Invention of a Tool Which Cuts the Sod and Removes It In One Operation." *Boston Globe*, May 30, 1909, 40. At West Island in Fairhaven, MA, "it has been infested with the salt marsh mosquito, the fiercest of all the mosquito tribe. … Ditching is the very simple means employed to destroy the breeding places. … When the public is positively assured that the work on West Island has proved effective there can be no doubt but that this method will be employed in the development of other shore tracts throughout New England." Traces history of ditching in New Jersey. Discusses invention of a shovel to remove soil down to more than 25 inches, claiming "To leave the roots [of the wild grasses] in the ditches would be distinctly a bad feature, for in growing the roots would soon fill the ditches, thus stopping the flow of the water, making the water stagnant and inviting the mosquitoes there to breed." The West Island project, first of its kind in Massachusetts, involved 40,000 feet of ditches, with the main ditch, 20 inches wide, running parallel with the shore.

e. "Winning Hard Mosquito War; Pest Vanishing at Revere. Dr Lamb, Specialist on Subject, Being Thanked. Conditions Made Problem Difficult One." *Boston Globe*, August 10, 1912, 9. Dr. Lamb was the inspiration behind the "ditching plan" for some 1,100 acres of marsh. "Revere is the literal mosquito headquarters of New England," this at a time when there were "150,000 to 200,000 persons who frequent the beach on the hot days." Revere Beach, opened July 12, 1896, was America's first public beach. Katie Redefer, "Revere Beach marks 125 years as the nation's first," *Boston Globe*, July 14, 2021, B3; "Revere Beach Reservation Historic District," National Park Service website, https://www.nps.gov/places/revere-beach-reservation-historic-district.htm.

f. "Medford," *Boston Globe*, July 5, 1913, 4. "Work ditching the marsh and swampy land along the Mystic River banks below Cradock Bridge for the purpose of wiping out the stagnant water pools which form mosquito breeding spots is expected to begin next week [at a cost of] $15 an acre. There are about 525 acres to be ditched. The cost will be borne by the owners of the property and will be levied as a betterment."

g. "Scituate," *Boston Globe*, September 21, 1931, 11. "The State Commission of the South Shore mosquito control project has closed its headquarters at Egypt [a section of Scituate]. It was organized with Selectman James W. Turner chairman, and Henry Webb superintendent. Extensive ditching of marsh lands was carried on during the Summer in Cohasset, Scituate, Marshfield, Duxbury and Plymouth to eradicate the mosquitoes. Two hundred men were employed."

7 Mosquito Ditches

History of mosquito ditches — OWMMS, App. A, cited above. In addition, studies of modern salt marsh management.

a. The Norfolk County [MA] Mosquito Control District, "A Brief History of Mosquito Control and How the Norfolk County Mosquito Control District Came To Be," NCMCD website, https://norfolkcountymosquito.org/about/. Excellent historical survey of mosquito control.

b. Ben Palmer, "Swamp Land Drainage with Special Reference to Minnesota," *Studies in the Social Sciences, no. 5, Bulletin of University of Minnesota* (Minneapolis: U of MN, 1915), http://www.minnesotalegalhistoryproject.org/assets/Palmer%20on%20Drainage%20(1915).pdf. ("Swamp" includes marshes.) Exhaustive analysis of drainage legislation (see App. 1 Drainage statutes of states other than Minnesota), history of drainage (going back to Caesar), and need for drainage to improve health.

   i. "It has been estimated that there are in the United States to-day approximately 80,000,000 acres of swamp and overflowed lands, an area of unproductive land …" Page 1.

   ii. "The decrease in malarial diseases as a result of the drainage of swamp lands is well known." 2.

   iii. "The reclamation of these swamp lands will include three distinct operations: (1) The construction of ditches, or the improvement of natural drainage channels, such as will be required by a large number of land owners in common, and will necessitate the use of the power of the state and of legal procedure; …" 3.

   iv. "In early times, when the Netherlands were sparsely inhabited by uncivilized tribes, the country was a vast morass periodically inundated by the waters of the North Sea. A few years after the territory had been subjugated by Julius Caesar, Pliny wrote of the land as follows: "'There the ocean pours in its flood twice a day, and produces a perpetual uncertainty whether the country may be considered as part of the continent or of the sea. The wretched inhabitants take refuge on the sandhills, or in little huts which they construct on the summits of lofty stakes whose elevation is conformable to that of the highest tides.'" 5.

c. A. Glenn Richards, Jr., "Mosquitoes and Mosquito Control on Long Island, New York, with Particular Reference to the Salt Marsh Problem" in *New York State Museum Bulletin, No. 316* (Albany: University of the State of New York, September 1938), https://www.biodiversitylibrary.org/item/228653#page/1/mode/1up.

Mentions early antimosquito efforts in 1900. Mosquito ditching began by 1916. In 1934–1936 WPA labor was used. Work moved from "checkerboard" ditches to parallel ditches, with some of the old cross ditches being filled in. "Actually this reduction in malaria on Long Island can be due only in part to the activities of the mosquito commission."

d. Alfred C. Redfield, "Development of a New England Salt Marsh," *Ecological Monographs*, vol. 42, no. 2 (Washington, DC: Ecological Society of America, Spring 1972), https://doi.org/10.2307/1942263. See 226-230 (mosquito ditches cause overdrainage of the marsh).
e. Various Google Books Ngram Viewer searches for "ditch,ditching" (from 1800 on, but you can specify earlier years), including
    i. Example from 1803 of *Private and Special Statutes of the Commonwealth of Massachusetts*, in vol. 3, page 159 et seq., "An Act to incorporate certain Proprietors of Salt Marsh, lying in Salisbury, in the County of Essex, to make and maintain a Dyke and Drains, for the better improving the same." March 5, 1803. "Sect. 3. *And be it further enacted*, That the Proprietors aforesaid be, and they hereby are authorized ane empowered to keep open and in good repair, a ditch which now runs easterly from the line drawn from land of *Nicholas French* to land of *Samuel Eaton* aforesaid, to a creek of sufficient width and depth to drain off the stagnant waters as aforesaid: *Provided nevertheless*, That the owners of the marsh through which the said ditch shall pass, shall be entitled to damages for any injury they may sustain by means of keeping the same open." https://books.google.com/books?id=XTY7XtgV-8MC&newbks=1&newbks_redir=0&source=gbs_navlinks_s.
    ii. Charles Lee Myers, "Status of Mosquito Control Work in Hudson County," New Jersey Mosquito Extermination Association, *Proceedings of the Annual Meeting of the New Jersey Mosquito Extermination* (1917), 44, https://books.google.com/books?id=8xI0AQAAMAAJ&source=gbs_navlinks_s: "On the elimination of the mosquito the future greatness of Hudson County stands. … We have made about 818,000 feet of ditches on our salt-marsh meadows. … The new ditches through the meadows formerly covered by forests where there remain any tree stumps were thirty inches wide. This is a great improvement in effectiveness over the standard ten-inch ditch …."
    iii. "The Story of the Mosquito," New Jersey Agricultural Experiment Station circular (New Brunswick, November 1921), 13 (with photos of hand and machine digging of ditches): "… the usual means of control are ditching, diking, and tide-gating, and filling of small areas, or large areas when feasible." Reprinted January 15, 1924, in "A Mosquito Manual" Circular 130,

(https://books.google.com/books?id=J0g2AQAAIAAJ&printsec=frontcover#v=onepage&q&f=false.

f. Plymouth County and state mosquito control agencies — see above
g. Town of Scituate annual reports. Recent years available from town website, older ones from Scituate Town Library website. See search for "ditch" at https://archive.org/details/scituatetownlibrary?query=ditch&sin=TXT.
    i. 1928. The 1928 annual report, page 182, describes the oiling of mosquito breeding grounds, then says, "The work as recommended by the State Engineer was carried out as long as the appropriation lasted. In all about two miles of ditching was done. Some of this work was done in the west end of the Town, some in the center of the Town and some at the Harbor. … There is a great deal of ditching to do in order to dry up the swamps and marshes, and by so doing, the work of eliminating the mosquito would be much easier." https://archive.org/details/annualreportofto1928scit/page/n423/mode/2up?q=ditching
    ii. 1931. Report of Selectmen, *Annual Report of the Officers and Committees of the Town of Scituate For the Year Ending December 31, 1931* (printed by Sanderson Brothers, North Abington, Massachusetts, 1932), 9–10, https://archive.org/details/annualreportofto1931scit/page/n25/mode/2up?q=mosquito. "In the early spring a campaign designed to rid the shore areas of Massachusetts of the mosquito nuisance was launched, and under recommendation of his Excellency, Governor Ely, the Legislature appropriated for the purpose the sum of $270,000, said sum to be expended under the direction of the State Reclamation Board. Four separate Mosquito Control districts were created by this Board, and to the South Shore Project, so-called, in which Scituate was included, was allotted the sum of $70,000. This was supplemented by appropriation from some of the Towns within the Project, the amount in the case of Scituate being $10,000 raised at the meeting of March 2nd, 1931. The actual operation of constructing drainage trenches started on April 6th on the marshes between Hollet Street and Cohasset Harbor, and there have been dug in the Town of Scituate alone over one hundred and fifty miles of trench, at an expense of about $25,000, and we are confident from the experience of the past season where marsh areas were drained that the results of this undertaking will be well worth the cost in making Scituate a more desirable place in which to live, and in enhanced values to our residential property, as well as affording a very substantial item of employment at a time when employment was appreciated." One hundred fifty miles is an astonishing amount. Later town reports never show more than one mile a year.

iii. Note: CCC established April 1933; WPA established April 1935.

iv. 1961, 126. "Throughout the months between past and coming mosquito breeding seasons much attention will be given to ditching. Work on the salt marsh will be done in the fall and early spring. … In addition to ground dusting, it is planned to continue application of D.D.T. insecticide dust by air …."
https://archive.org/details/annualreportofto1961scit/page/n133/mode/2up?q=ditching,

v. 1962, 99,
https://archive.org/details/annualreportofto1961scit/page/n387/mode/2up?q=sewer. "DITCHING

Drainage, always considered the keystone of mosquito control, is more than ever important as a result of a growing concern by certain groups over the use and buildup of insecticides and other chemicals in our wildlife areas.

Over the years the ditching done by this project has greatly reduced the areas that need spraying or dusting of insecticides. On fresh water only areas capable of mosquito breeding are drained, shallow stagnant pools. On salt marshes we attempt to keep all ditches clear of grass and mud to allow tidal water a free flow in and out, also to eliminate all brackish sheet water at upper ends of marsh where salt marsh mosquitoes will breed.

The largest amount of mosquito control can be reached with a good drainage program. With this idea in mind, we have proceeded in all areas possible to ditch. In the Town of Scituate 215 feet of ditches have been cleaned, 27,375 feet reclaimed and 1,200 feet of new ditch have been dug."

vi. 1968, 127. "Submitted herewith is the report of the South Shore Mosquito Control Project's activities for the year November 1, 1967 to October 31, 1968, … a year round cooperative effort embracing the City of Quincy and Towns of Braintree, Cohasset, Duxbury, Hingham, Hull, Marshfield, Norwell, Scituate and Weymouth, covering a total area of 172.21 square miles serving a population in excess of 223,000. … The Project is authorized and acts under provisions of Massachusetts General Laws, Chapter 252, Section 59, mosquito control provisions, and Chapter 112, Acts of 1931, pertaining to ditch maintenance of the salt marshes. … Source Reduction - By properly placed and maintained ditches on coastal marshes, tidal and storm waters are allowed to flow off, where otherwise in part would become brackish to breed mosquitoes. Fill from ditch excavating must be graded to insure water pockets are not left to breed."

vii. 1977 and 1980. The 1977 annual report, page 74, says "To flush or drain off stagnant water, 3600 feet of new ditch was constructed both in tidal and upland waterways. … To maintain ditch work previously constructed, 59,200 feet of marsh ditching was reclaimed." https://archive.org/details/annualreportofto1977scit/page/n377/mode/2up?q=ditching. Equivalent numbers in 1980 were 2450 and 110,300, https://archive.org/details/annualreportofto1980scit/page/n121/mode/2up?q=ditching.

viii. 1987–2008. In most of these years, the town attacked mosquito breeding grounds, including reconstructing salt marsh ditches and cutting new salt marsh ditches using a track driven excavator.

ix. 2013, 166–167. "The public health problem of EEE [Eastern Equine Encephalitis] and WNV [West Nile Virus] continues to ensure cooperation between the Plymouth County Mosquito Control Project, local Boards of Health, Massachusetts State Reclamation and Mosquito Control Board and the Massachusetts Department of Public Health. … Machine Reclamation. 450 linear feet of saltmarsh ditch was reconstructed in Scituate using the Project's track driven excavator."

x. 2020, 192. "Water Management: During 2020 crews removed blockages, brush and other obstructions from 1,585 linear feet of ditches and streams to prevent overflows or stagnation that can result in mosquito production. This work, together with machine reclamation [no amounts given], is most often carried out in the fall and winter."

h. "While mosquitoes in salt marsh are not known to carry either West Nile of [or] EEE, the SRMCB continues to maintain existing salt marsh ditches. These activities are exempt from the state wetlands protection act and local bylaws. Some tidally restricted marshes in Buzzards Bay identified in our salt marsh atlas were restored because the SRMCB removed fill that had accumulated in a salt marsh drainage ditch." "Mosquito Control around Buzzards Bay," Buzzards Bay National Estuary Program, https://buzzardsbay.org/buzzards-bay-pollution/phinfo/mosquito-control/.

8 Salt Marsh Modern Management, Threats, and Remediation

Much work and research is going into helping salt marshes withstand threats from past practices, as well as sea level rise.

a. David G. Casagrande, "The Full Circle: A Historical Context for Urban Salt Marsh Restoration," 13–18, in David G. Casagrande, *Restoration of an Urban Salt Marsh: An Interdisciplinary Approach, Bulletin 100* (New Haven: Yale School of Forestry and Environmental Studies, 1997),

https://elischolar.library.yale.edu/cgi/viewcontent.cgi?article=1101&context=yale_fes_bulletin.

b. Watson, E. B., Raposa, K. B., Carey, J. C., Wigand, C., & Warren, R. S. (2017), "Anthropocene survival of southern New England's salt marshes," *Estuaries and coasts: journal of the Estuarine Research Federation*, *40*(3), 617–625. https://www.ncbi.nlm.nih.gov/pmc/articles/PMC6161497/, and https://doi.org/10.1007/s12237-016-0166-1. "In southern New England, salt marshes are exceptionally vulnerable to the impacts of accelerated sea level rise."
c. "Restoring the Great Marsh," US Fish & Wildlife Service, July 17, 2014, https://usfwsnortheast.wordpress.com/2014/07/17/restoring-the-great-marsh/.
d. The Great Marsh, a 20,000-acre salt marsh in northeastern Massachusetts, is a crucial habitat for birds and aquatic species. Threats if marshes shrink or sink. See Mass Audubon, "Site Summary: Site Summary: Great Marsh," https://www.massaudubon.org/our-conservation-work/wildlife-research-conservation/bird-conservation-monitoring/massachusetts-important-bird-areas-iba/iba-sites/great-marsh.
e. Quote from Historic Ipswich [MA] website, https://historicipswich.org/environment/:

   The Great Marsh restoration project: Marshes serve as important habitats for sea life that support the local ecosystem and seafood economy and provide a natural flood barrier to protect neighboring communities. The project aims to fortify 300 acres along Old Town Hill and two other Trustees sites in Essex and Ipswich. Over time historic ditching processes have compromised the resilience of the marsh by destroying its natural draining process, leaving it increasingly vulnerable to floods. In order to 'heal' these ditches, the Trustees and partners will use an innovative, nature-based method of "ditch remediation"
f. Caitlin Mullan Crain, Keryn Bromberg Gedan, Michele Dionne, "Tidal Restrictions and Mosquito Ditching in New England Marshes: Case Studies of the Biotic Evidence, Physical Extent, and Potential for Restoration of Altered Tidal Hydrology" in Pt. 3: Land Use and Climate Change, in Brian R. Silliman, Edwin D. Grosholz, Mark D. Bertness, editors, *Human Impacts On Salt Marshes: A Global Perspective* (Berkeley: University of California Press, c.2009), article at http://www.edc.uri.edu/nrs/classes/nrs555/assets/readings_2011/Crain_etal_2009_Chap9.pdf. Good review, great bibliography.
g. David M. Burdick, Gregg E. Moore, Susan C. Adamowicz, Geoffrey M. Wilson, and Chris R. Peter, "Mitigating the Legacy Effects of Ditching in a New England Salt Marsh," *Estuaries and coasts* 43, no. 7 (2020), 1672–1679, doi: 10.1007/s12237-019-00656-5.

h. NSRWA talk, Feb. 2022, cited above — sparrows, spartina, height/sinking of marsh, effect of ditching, process of ditch remediation.
i. Dustin Luca, "Saving the salt marsh: Volunteers undoing centuries of human activity," *The Salem News*, May 26, 2022, https://www.salemnews.com/news/saving-the-salt-marsh-volunteers-undoing-centuries-of-human-activity/article_b3f7a80c-db8f-11ec-a961-0b2a88ab107d.html.
j. Sarah Shemkus, "Trustees Work to Restore and Preserve the Ailing Great Marsh," *NorthShore* magazine, October 5, 2020, https://www.nshoremag.com/faces-places/trustees-work-to-restore-and-preserve-the-ailing-great-marsh/.
k. 2016 conference: "Susan [Adamowicz]: "Since Hurricane Sandy, we've come to understand that barrier islands and coastal marshes are naturally designed to be resilient and to protect our shores. Continuing to remove prior alterations and restore more natural processes such as tidal flow, sedimentation and healthy vegetation will help maintain good quality marshes. Where coastal systems have been highly altered, we see the need for larger/more extensive resiliency efforts. … Just as we, as a nation, are turning to the restoration of our built infrastructure (roads and bridges), we also need to be mindful of the natural infrastructure — coastal wetlands and barrier islands — that protect our shores and coastal communities." https://medium.com/usfishandwildlifeservicenortheast/highlights-from-americas-biggest-conference-on-coastal-restoration-6a89a590f8a6.
l. Unknown author, slide presentation, 10/26/2016, "Impact of Mosquito Ditching on the Spatial Distribution of Salt Marsh Vegetation," https://web.uri.edu/ltrs/files/Impact-of-Mosquito-Ditching.pdf.
m. Marc Carullo, Massachusetts Office of Coastal Zone Management, "Marsh Impairment and Future Considerations: A Massachusetts Overview," NERRS/NEERS Salt Marsh Special Symposium, Portsmouth, NH, April 26, 2018, http://nbnerr.org/wp-content/uploads/2018/09/Marsh_impairment_and_future_considerations_-_a_Massachusetts_overview_M._Carullo.pdf. Photos show loss of marshland.
n. Eastern Research Group, *Report for the Mosquito Control for the Twenty-First Century Task Force* (August 2021), 216, https://www.mass.gov/doc/mosquito-control-task-force-report-august-2021/download:
"6.4.2 Ditch Maintenance
Ditching, a formerly widespread method of mosquito control, disrupts natural water flow and topography with artificial ditches and channels. This method is no longer common in New England. However, these states still maintain some ditch systems and channels … Research on the effectiveness of ditch maintenance for mosquito control has indicated that the method is only

effective if water is kept at least 1 meter deep. Walton (2011) notes that ditches and water channels that are kept deeper than 1 meter form deep-water zones and reduce the amount of standing, shallow water for mosquito breeding. Deep water also allows fish and other mosquito predators to move through different areas and access mosquito larvae and egg rafts."

o. Eastern Research Group, "Final Report: Mosquito Control Task Force Study" (September 2, 2021), presentation, https://www.mass.gov/doc/mctf-study-final-report-presentation/download.

p. In Massachusetts, Mosquito Control Districts (MCDs) file annual reports with the State Reclamation and Mosquito Control Board, and they are available online here: https://www.mass.gov/service-details/state-reclamation-and-mosquito-control-board-annual-reports. The website of the Plymouth County Mosquito Control Project has its 2021 annual report, along with individual reports summarizing activities in each town, which towns include in their annual reports. https://www.plymouthmosquito.org/annual-reports.html. The Plymouth County 2016 report, page 8, said: (1) 10,852 feet of saltmarsh ditches were maintained, and (2) OMWM was active year round. The county's 2021 report, page 8, said: (1) only 1,090 feet of saltmarsh ditches were maintained, and (2) Open Marsh Water Management (OMWM) "is not in use due to current restrictive regulations as well as possible negative impacts to the salt marsh when combined with sea level rise."

q. The Marsh Sustainability and Hydrology Project, sponsored by the Woods Hole Oceanographic Institution, brings scientists and end-user organizations together to study marshes. https://www2.whoi.edu/site/marshsustainabilityandhydrology/#home-about-us. It has developed a modeling tool for coastal planning and wetland management. Its work is described at the website of the Friends of Barnstable Harbor, "Salt Marshes: Ditches, Drainages, & Sea Level Rise," https://www.friendsofbarnstableharbor.org/resources-information/salt-marshes/. The website describes a study that discovered that mosquito ditches accelerate drainage, causing sediment to dry out, "no longer forming the waterlogged peat that can preserve roots and other organic components for hundreds of years. The result is erosion and lower marsh elevations. In the face of predicted sea level rise, this could lead to the 'drowning' of the marsh." (underlining omitted).

r. Maher, N., Salazar, C. & Fournier, A., "Advancing salt marsh restoration for coastal resilience: a learning exchange," *Wetlands Ecol Manage* (2021), https://link.springer.com/article/10.1007/s11273-021-09841-5, and https://doi.org/10.1007/s11273-021-09841-5. Interesting, some useful insights. Runnels (narrow channels) were incorporated in marshes to add sinuosity.

s. "How sea level rise is affecting our coastal salt marshes and what we can do about it," Massachusetts Land Conservation Conference, March 26, 2022, https://massland.org/sites/default/files/files/seal_level_rise_what_to_do_mar26_9a.pdf. Three presentations at 2022 conference on Sea Level Rise effect on salt marshes. Many charts, nice photos. MA has 45,000 acres of salt marsh.
t. Jon Woodruff, Hannah Baranes, et al, "Understanding Sediment Availability to Reduce Tidal Marsh Vulnerability to Sea Level Rise in the Northeast," UMass Amherst website, https://necasc.umass.edu/projects/understanding-sediment-availability-reduce-tidal-marsh-vulnerability-sea-level-rise; UMass Amherst, "How a Massachusetts salt marsh is changing what we know about New England's coast," Science X website, March 14, 2022, https://phys.org/news/2022-03-massachusetts-salt-marsh-england-coast.html. And *ScienceDaily*, March 14, 2022, www.sciencedaily.com/releases/2022/03/220314154417.htm.
u. David Abel, "Blocking the Herring River a century ago brought disastrous ecological consequences. Now, officials want to resurrect it." *Boston Globe*, updated July 27, 2022, https://www.bostonglobe.com/2022/07/27/science/century-later-resurrecting-river-vital-ecosystem/. Also in print edition (as "Going with the flow once more"), July 28, 2022, 1. Remediation in Wellfleet, MA. Note reader comments on article at *Globe* website.
v. Roy Schiff, et al, to Town of Hampton Public Works Department, "Site Assessment and Selection Summer – Ditch Remediation in the Hampton Salt Marsh," letter, August 10, 2022, http://www.hamptonnh.gov/documentcenter/view/5485.

9 Questions for Further Research

a. What are those ditches in the salt marshes?
b. Do we need to maintain them to control mosquito-borne diseases? Do we need to maintain *all* of the ditches in a marsh to control the diseases?
c. How are salt marshes threatened? Sea level rise, sinking marshes, perhaps due to ditches (erosion of sediment at low end of roots; will addition of sediment on top help?); harm to sparrows
d. Can ocean deposition of sediment help? See UMass Amherst research.
e. How does salt marsh remediation work?
f. Does filling in ditches (remediation) conflict with public health (maintenance of mosquito ditches)? Or complement it? Any middle ground? What about "thin-layer deposition of dredged material to enhance tidal marshes that are suffering elevation deficits …"?
g. Are all marshes sinking? What about those along North River (Third Cliff)?

h. Should we monitor the height of our marshes? Who should do this? What (if any) past data?
i. What material is available for further research? What more needs to be done? WPA or state agency records? Massive, perhaps unhelpful (hours worked) but maybe photos and plans. Also, local histories, pretty obscure or difficult to find. A few are mentioned above, including Vinal's *Salt Haying in the North River Valley (1648–1898).*
j. Consider peatland under (or part of) marshes? Howe's *Salt Rivers* describes well.
k. Consider dikes/dykes? Dams? Embankments (banks, or berms) created when ditching? See discussion in NSRWA talk.
l. Consider ditching in cranberry bogs? (They tend to be inland.) See Kimberly Sebold, "From Marsh to Farm," ch. 7. They were ditched as early as the second half of the 1800s; "Cranberries," *Boston Globe*, September 3, 1874, 2.
m. Graph amount of mosquito ditching added each year (by town, region, state)?
n. Consider marshes as carbon sinks?
    i. Tatum McConnell, "Protecting nature's best carbon sink: Peatlands," Scienceline, January 7, 2022, Scienceline, https://scienceline.org/2022/01/protecting-natures-best-carbon-sink-peatlands/. "Rising temperatures and development threaten Earth's peat; without it, our climate would warm even faster."
    ii. Kimbra Cutlip, "For the World's Wetlands, It May Be Sink or Swim. Here's Why It Matters," *Smithsonian Magazine*, January 13, 2016, https://www.smithsonianmag.com/smithsonian-institution/worlds-wetlands-it-may-be-sink-or-swim-heres-why-it-matters-180957808/. "Despite taking up just four to six percent of the Earth's land area, wetlands such as marshes, bogs and mangrove forests hold a quarter of all the carbon stored in the Earth's soil."

I close by turning to the West Coast, and recommending a great recent article. It illustrates human uses of a coastal environment, along with recent attempts to restore it. That environment consists of salt flats in tidal wetlands at South Bay, near San Francisco. This is where salt was cultivated and harvested, by indigenous persons, and by industrial corporations with dikes. Their impact is wonderfully described and colorfully illustrated by Skylar Knight (text) and joSon (photos), "Past the Salt," *bioGraphic*, July 14, 2022, https://www.biographic.com/past-the-salt/, republished as "Rebirth of San Francisco's Salt Marshes," *Hakai Magazine*, August 16, 2022, https://hakaimagazine.com/features/rebirth-of-san-franciscos-salt-marshes/.

Back to Introduction

About the Author

Author photo by Kjeld Mahoney

Lyle Nyberg lives near salt marshes in Scituate, Massachusetts. One of his English ancestors settled in Scituate in the 1630s and built the 38th house in town. There is no evidence that he ditched the marshes.

Lyle graduated from Dartmouth College and Boston University School of Law. He is a lawyer turned independent scholar and historian. Before researching ditches, he wrote and published three books on historical topics. He documented more than 50 historical buildings in the greater Boston area, especially those relating to women's suffrage.

He is a member of the Scituate Historical Society, the Massachusetts Historical Society, and the New England Historical Association.

He can be reached at www.lylenyberg.com.

Made in the USA
Middletown, DE
16 September 2022

10515887R00024